I0016641

The Fundamentals of SEO for the Average Joe

An in-depth guide for Search Engine Optimization (SEO) and the basics of Search Engine Marketing (SEM) simple enough for any Average Joe.

Written by Alyssa Ast
Edited by Angela Atkinson

ISBN:978-0-557-59512-9

The Fundamentals of SEO for the Average Joe

A basic guide for all aspiring and seasoned freelance writers interested in learning the art of increasing web traffic using search engine optimization and search engine marketing.

The Fundamentals of SEO for the Average Joe

Table of Contents

The Fundamentals of SEO for the Average Joe

Part 1:
Search Engine Optimization

Introduction to Search Engine Optimization

Search engine optimization (SEO) is an essential skill for anyone seeking to profit in the online content world. Acquiring SEO skills can offer freelance writers the opportunity to increase their income as the world becomes increasingly reliant on the Internet for their information and communication. Plus, by employing their SEO skills in their own work and on their author sites, writers are afforded the opportunity to increase their profit margin through increased web traffic. Ultimately, increasing traffic to online content offers various benefits to clients and writers alike.

Benefits of Search Engine Optimization:

SEO knowledge can add spark to any resume for the simple fact that increased web traffic results in a profit. As the world continues to revolve around the Internet, it is vital that online content is ranked as high as possible on a *search engines results page (SERP)* to increase content visibility. Clients seek writers and SEO experts who can accomplish this goal, since SEO help can attract more website visitors and increase company visibility, thus increasing potential profits.

For writers, not only is SEO a critical resume booster, but it can improve their business prospects as well. Increasing traffic to your online content, such as articles, websites, and blogs, will generate a strong readership and client potential. Plus, with proper content optimization, a large profit can be generated from *pay per views (ppv's)* through some content sites. Profit can also be generated from clients that would not have found your writing services without an ideal SERP ranking.

The internet has become a permanent tool for everyday life, and so has search engine optimization --although some consider it to be a dying trend. But in truth, SEO continues to grow and evolve, just as the internet continues to do so. As a result, the basic fundamentals of SEO are here to stay. For that reason, the sooner this vital skill is acquired, the greater your outlook as an online content producer becomes.

What is Search Engine Optimization?:

Despite what you may have heard, search engine optimization is actually an easy skill to learn. Basically, SEO uses the strategic placing of *keywords* and phrases to increase the visibility of a website on the results pages (SERP's) of search engines like Google and Yahoo. By utilizing SEO tactics, web content stands an increased probability of accumulating traffic than if the natural approach without optimization had been taken.

SEO involves taking into consideration how people search for information on the internet. By understanding the most common methods used to find information online, you can then use those methods to increase web content traffic. As a result, this creates *"search engine friendly"* content.

As the Internet evolved, so did the techniques used for search engine optimization. These days, *search engine marketing (SEM)* and search engine optimization are often employed concurrently. By combining these two methods, the potential for traffic significantly increases. The combination of the two methods is often referred to as *Search engine marketing professional organization (SEMPO),* which is believed to be the wave of the future.

The focus of '*The Fundamentals of SEO for the Average Joe*' is that of search engine optimization. However, search engine marketing will also be briefly covered, including the use of *hyperlinks* and *backlinks.* The SEM portion of this guide will discuss how to make money with SEM and how to market yourself using this skill.

Utilizing Keywords and Key Phrases

Utilizing the correct keywords and *key phrases* is the single most important aspect of SEO. Keywords and phrases are even more important than keyword density. The terms "keywords" and "key phrases" refer to the words or phrases people type into a search engine to find information. For example, if someone is researching how to make homemade pies, they might type in "homemade pie recipes" or "how to make pies." These are key phrases. A keyword for the topic would be "pie" or "homemade." To optimize content, your goal is to find the best keywords and phrases that relate to the web content you're optimizing. However, sometimes you may need to get a little creative with them depending on the *hits number*.

How to Find Keywords and Phrases:

You can find vital keywords and key phrases through the use of a keyword tracking tool. There are many tools available online to help you choose the words and phrases that are ideal for your content. One free and easy-to-use keyword tracking tool can be found atWordTracker.com.

When using a keyword tool, you will type a word or phrase that relates to your content and the tool will provide you with a list of similar words and phrases, as well as the number of times *(hits number)* that word or phrase is typed into a search engine per day.

Sample :

Search Phrase: Sick Cats

Keyword	Search Results	Hits #
1	Sick cats	36
2	Signs of sick cats	14
3	Sick cats with fleas	3
4	Sick cats indoors	2
5	Sick cats and hissing	2

Despite what many people think, you do not want to choose words and phrases that have high hits' numbers. You want avoid using these words because the higher the hit number is, the higher your competition will be.

Your goal is to find the keyword or phrase that has a low hits number and one that fit's with the content you are optimizing. It is best to choose a word or phrase that has less than 20 hits per day. The ideal word or phrase would have around 10 or less hits per day. From the example above, the best keywords and phrases to use would be: sick cats with fleas, sick cats indoors, or sick cats and hissing.

Sometimes finding words and phrases with a low number of hits can be difficult. This is when you must get creative with the words and phrases you search. If you still cannot find a word or phrase to fit your needs, you can choose a keyword with a higher hits number; but, never one with over 50 hits. The words and phrases you decide to use must be used within the content exactly as they appear on the word tracker tool in order for optimization to be successful.

An essential aspect of using keywords and phrases is updating. To keep traffic flowing to web content and to maintain high SERP's ranking, keywords and phrases should be updated at least once per month. The content people search for is changing everyday; therefore, the hits number is constantly changing as well. For this reason, updating is vital for maintaining optimization success.

Primary and Secondary Words and Phrases:

To most effectively optimize web content, you will need at least a primary and secondary keyword or phrase. As we explore *keyword density*, we will cover the use of up to five keywords or phrases, but for now, we'll focus on primary and secondary. You will use the primary and secondary within your title, *Meta description*, web content, and *Meta tags* section. Meta descriptions and Meta tags will be covered in detail later in this book.

When choosing your primary and secondary words or phrases, pick one keyword or phrase that is simple and consists of one or two words. You will also want to pick a *long tail keyword phrase.* This is a phrase that consists of three to five words. It is best to make your *secondary keyword or phrase* the long tail key phrase.

Optimization: The First Steps

Creating optimized titles for content, such as articles and blogs, requires using different techniques than those used to optimize web pages for websites. In this section, we will cover how to optimize the titles used for web content and websites, as well as utilizing Meta descriptions and Meta tags.

Optimizing Content Titles:

Optimizing content titles requires a bit more thought than simply writing a title. Keywords and phrases to optimize content must be used correctly within the title to produce benefits. Through combining properly placed keywords and phrases, along with attention grabbing words, the probability of a reader choosing your content over competing content is significantly increased.

After you have found your primary and secondary keyword and phrase to use as a title, seek out eye-catching words to capture a reader's attention and entice him to click on your content. There are certain words that are known to cause a reader to have an emotional reaction that captures their attention.

Here are some examples of those words:

Free	Danger
Best of	Avoid
Top 10	Win

According to SEO researchers, these kinds of words catch a reader by surprise and draw them in purely out of curiosity--boosting your content traffic. When creating the title, keep the attention grabbing word at the beginning of the title whenever possible.

The keywords and phrases you have chosen for your title should be brief and to the point to avoid confusion and title stuffing. When using the keywords and phrases, never place them right next to each other. Your title should flow naturally, but space out the keywords. Try to have at least two words between each keyword/phrase in the title.

Here's an example:
Primary: Publishing Information
Secondary: Freelance Writers

*"The Best **Publishing Information** for Online **Freelance Writers**"*

After you know what you're doing, creating optimized titles is easy. However, there are common mistakes many people make that hurt traffic levels. Here are some common mistakes to avoid:

- Create short titles. Titles that are too lengthy-will bore readers and decrease SERP's ranking.
- Create unique titles. Duplicating titles for your content (even on different sites) your work to compete with itself and lowers SERP's ranking.

- Do not stuff titles with keywords. The titles should flow naturally to eliminate confusion.

Optimizing Website Titles:

As mentioned previously, optimizing website titles is different than optimizing content titles. When optimizing website titles, you will only want to use one keyword or phrase and it should be as short and simple as possible. While you will still need a secondary phrase for the Meta description, only use your *primary keyword/phrase* within the title. It's essential the primary phrase you choose is as direct as possible for the sake of the webpage's URL.

The vast majority of websites create a webpage's URL from the title provided on the webpage. This is why it's essential that your primary phrase is as simple as possible. Longer URL's tend to have a lower SERP's ranking than those that are shorter. Using your primary keyword/phrase, create a simple and direct title for the webpage's title.

Optimizing Meta Descriptions:

Meta descriptions play a vital role in your content's SERP's ranking, although it is not seen within the content. Meta descriptions are generally about 200 characters long, including at least the primary and secondary keywords and phrases. While the Meta description is not visible on a webpage, it is the information that appears on a SERP's page. A Meta description actually appears as the little section on the SERP's page that gives a bit of a teaser about the information provided within a webpage.

Proper optimization of Meta descriptions is essential for SEO success. The Meta description optimization process for content and websites is the same. Because you are only offered limited space for a Meta description, you have to creatively make the most of it. Incorporate your primary and secondary keywords/phrases within the description while enticing readers to choose your content over others.

The content must flow naturally and be clearly written and easy to understand, while encouraging readers to pick your content. Using the previously mentioned example key phrases-, here is an illustration of a Meta description:

Primary: Publishing Information
Secondary: Freelance Writers

"Freelance writers are often bombarded with publishing information that is of little value. Luckily, there an ideal publishing resource that suites everyone's needs."

Use each of your keywords/phrases one time within the Meta description. Once you understand keyword density, you can begin incorporating other keywords/phrases besides your primary and secondary into a Meta description.

Utilizing Meta Tags:

To successfully optimize content, utilize the *Meta tags* as well. Meta tags provide search engines with valuable information needed for your content's SERP's ranking. Basically, these tags are the keywords and phrases you have chosen for your content.

In the Meta tags section, always include your primary and secondary keywords/phrases. Don't forget to utilize the tags section because without it, your content won't stand a chance on a SERP's ranking. When you begin experimenting with keyword density, you can incorporate a few more keywords into the Meta tags section.

Optimizing Web Content

Optimizing web content also revolves around using the proper *keyword density* level to increase traffic. Keyword density is simply the percentage of keywords and phrases used within web content compared to the full content's *word count*. It is vital to use proper keyword density because if your density is too low, your SERP's ranking decreases. But, if your density is too high, a search engine will flag it as spam. All the rules of keyword density apply to all forms of online content.

Keyword Density:

There is much debate over the correct percentage to use for keyword density. It commonly ranges from 2% to 8%, but sometimes is said to be as high as 12%. Certain clients will request a percentage to use when optimizing their content. Most often, I use a keyword density level of 4% for content ranging around 600 to 1,000 words. I use a smaller density level of 2% or 3% for shorter content.

For longer content, I often use up to five keywords or phrases to maximize traffic. For shorter content, I typically stick with 2 to 3 keywords/phrases. When using more than 2 keywords remember to include them in the Meta tags and Meta description.

Keyword Density within Content:

Determining how many times your keywords/phrases should appear in your content involves simple math. Take the word count of your content and multiply it by the keyword density you decide to use.

Here are a few examples:

750 words x 4% (.04) = 30
375 words x 2% (.02) = approx. 8
500 words x 3% (.03) = 15

When you begin writing your content, make sure it flows naturally and does not seem forced. All keywords and phrases have to be used exactly as they appear in the word tracker tool. This is important because if they are not used properly, your SERP's level will decrease.

Your primary keyword or phrase needs to be included within the first paragraph, preferably within the first sentence. Try to place the keywords and phrases evenly throughout the content, while maintaining the natural flow and avoiding stuffing the article.

Determining how many times you use each keyword and phrase within a piece of content involves basic math, along with making educated choices. Your primary and secondary keywords/phrases need to be used the most, whereas any other keywords will be used less often.

Here is a breakdown of how I typically determine how many times to use each word and phrase:

600 words x .04 = 24

Primary: Professional Information- 9 x's
Secondary: Freelance Writers- 6 x's
Keyword: Journalism- 3 x's
Key phrase: Breaking News- 3 x's
Key phrase: Writing Tips- 3 x's

There are many free tools available online to check keyword density levels of online content. One such tool that's quite effective is SEOtool.com. This online tool provides you with a full breakdown of every word and phrase used within the content. This allows you to easily see where you need to make adjustments to promote the most traffic to the content.

Maintaining Optimization:

As stated earlier, keywords and phrases need to be updated at least once per month to maintain a steady flow of traffic and a high SERP's ranking. This may also include adjusting the keyword density at times.

Freshly updated content always results in higher SERP's ranking. Websites often lose SERP's ranking because they are not updated often enough. One way to avoid this is to include a blog on the website. Regularly posting on the blog and incorporating SEO within post content instantly boosts a website's SERP's ranking.

Part 2:
Search Engine Marketing

Introduction to Search Engine Marketing

Search engine marketing (SEM) is a marketing method used to increase web content traffic and to generate a profit. SEM involves incorporating links within web content to advertisements and other sites to boost traffic and profit potential. SEO is the foundation of SEM. Content must be search engine optimized prior to incorporating SEM methods or SEM is not possible.

The keywords and phrases used for SEO are needed to apply links for SEM results. You must understand hyperlinking in order to use SEM. Utilizing all available methods for generating inbound and outbound links is essential for creating successful SEM and SEO content.

Hyperlinks:

Hyperlinks are an indispensable aspect of SEM. A hyperlink is a network link that connects one webpage to another. Hyperlinks are easy to recognize. They look like an ordinary word, except that they're usually underlined and often a shade of blue, but can appear in other colors as well. When you run the mouse curser over a hyperlink, a small hand will appear, allowing you to click on the link. A URL will also appear as the curser is placed over the link. There are many forms of links that can be used as a hyperlink, which we will discuss later.

There are two methods that are typically used to hyperlink content. One method involves using a tool provided on most content building websites and word processors. The other method involves inserting HTML code.

The first method we will discuss is the easiest and is used most often. On many content building websites you will see a button that looks like a small blue ball (similar to a globe) or an icon that looks like chain links. On word processors, if you right click on the screen, this will appear on the bottom of the option box that appears.

To hyperlink a word, follow these simple steps:

1. Type the word you wish to hyperlink, such as *'Making Money.'*
2. Highlight the word you want to hyperlink by either right clicking the mouse and dragging until the word is highlighted or use the "shift" and arrow keys on the keyboard to highlight the word.
3. Next, select the hyperlink option we discussed previously. A small box should appear that allows you to enter a URL.
4. In the box that has appeared, enter the URL of the page you want to link the original content to and click 'ok.'
5. The word should now be underlined and a shade of blue. Click on the link to ensure it opens to the correct page.

The second method that is used to hyperlink requires remembering an HTML code. The code is as follows:

<a href="<u>URL</u>">**Link text**</a>

The underlined section, <u>URL</u>, on the HTML code above is where you will insert your URL link for the page you are connecting to the original content. The bold words 'link text' is where you will insert the word(s) you want the link to be anchored to. Here's an example of how the code should look with the URL and text inserted:

<a href="http://www.alyssaast.com/">AlyssaAst</a>

                ^   ^

         <u>URL</u>   **Link Text**

With this code inserted, the hyperlink should appear the same as with using the hyperlink option on a webpage or word processor-- an underlined word that is generally bluish in color. Again, check the link to ensure it opens to the correct page.

Once you have hyperlinked web content to one another, link can be edited or removed if you like. To remove the link it is as simple as deleting the hyperlinked word and retyping it. Some content management systems will have a feature to remove the link as well. After editing a link always, always make sure it works properly and opens to the correct webpage.

Types of Links:

There are two types of links that are used for SEM: anchor links and naked links. The links discussed in the hyperlink section are *anchor links*. Anchor links are links that are anchored into text. These are considered to be hidden, other than the underlining of the text the link is anchored to.

Naked links are just as they sound. They are not linked to anything. Naked links are simply a URL inserted into text. Here's an example of a naked link:

http://www.alyssaast.com/

For SEM, anchor links are used most often but there are times that may call for naked links. It will also involve your personal preference when deciding which form to use

Outbound Links:

Outbound links are essential for supporting web content with third party websites to provide relevance to your content. They also allow search engines to index your content. Content is considered to be more reliable if outbound links are included. When creating outbound links, there are some important tips to remember.

Only link to sites that are relevant to your content. Link to *authoritative websites* only, not *user generated* content. Authoritive websites give your content more credibility. Authoritative sites typically include .gov, .org, and .edu at the end of the URL.

Make sure the text your link is anchored to makes sense. The link text should be relevant to the third party site you are linking to and should maintain the flow of the content.

Have your outbound links open to a new window. This will keep your content from essentially getting lost and the reader is more likely to refer back to your content. This is important for your *bounce rate.*

Utilizing Search Engine Marketing

SEM is used for various reasons. Most often, SEM is used to generate a profit with pay per click methods and to market a service or product with increased web traffic. Combining SEO with SEM is the ideal technique to market yourself as a writer and to boost your content traffic. All of the methods discussed in this section can be used in combination with one another to increase traffic and profit potential.

Making Money with Pay Per Clicks:

Incorporating *pay per clicks* into content is a great method to use for making money with search engine marketing. Pay per clicks use advertisements provided by advertisement companies. These advertisement links are incorporated into web content in the form of anchor links to entice readers to click on the link. Each click generates a small profit.

In order to profit from pay per clicks, the content must first be optimized and the links must flow with the content. The links must also be relevant to the content on the webpage. The goal is to hide the links as best as possible so they appear to be other web content and not advertisements.

Link Building and Backlinks:

Link building using backlinks is an ideal technique to market yourself and to increase website traffic. Link building involves using *backlinks* to create inbound traffic to web content. This is especially helpful if pay is received for page views. *Inbound* links also boost websites traffic, increasing a website's SERP's level.

One method used for link building is utilizing *reciprocal links.* Reciprocal links build website traffic and a third party's website traffic as well. Using reciprocal links involves the swapping of links to one another's website. It is a win/win situation for all involved. In involves finding a website with content similar to yours and agreeing to anchor their URL in your content if they return the favor. It is kind of like a "you scratch my back, I will scratch yours" situation. It is a great way to receive an inbound link, as long as both linking sites relate to one another.

Another method used to link build for inbound links it simply *posting for links.* All this involves is posting your URL wherever and whenever possible. This can include social networking sites, forums, and ads. Basically, anywhere and everywhere you can post a link, you post one. It works best if the links are posted in places that are relevant to your

site. You can also post links in comment sections of blogs and other sites. This will entice readers of that site to follow your link, increasing website traffic.

You can also buy links by participating in a buying links program. This is probably the least effective method used to link build because search engines often classify this content as spam or flag it as a potential virus.

The final method used to increase traffic with link building is to attract natural links. This is the best method to generate traffic to a site. *Attracting natural links* involves creating great content on a website and optimizing it. People love great websites and often spread the word about sites they favor. By taking the time to create a great site that is of use to others, you are likely to always have a steady flow of traffic. Place useful information on your site, and combined with a steady flow of traffic, readers are more likely to share the link with others, link your site to theirs, and visit your site frequently. This, along with SEO, is the best possible way to boost your website traffic and pay per view potential.

Glossary

Anchor Links- An URL link that can be found within content text, which links to another web page. Often used for SEM and increasing website traffic.

Attracting Natural Links- The act of attracting inbound links to web content with SEO and SEM.

Authoritative Websites- Credible websites that are not user generated. Typically include: gov, org, or .edu in the URL.

Backlinks- Links from a third party website linking to the original website, generating increased website traffic.

Bounce Rate- Used for web analysis to determine the amount of website traffic that either bounces from one website to another and the traffic that remains on a web page for a certain period of time.

Hits Number- The amount of times a keyword or phrase it typed into a search engine per day.

Hyperlinks- Links used to connect web pages to one another.

Inbound Links- External links to your site on other websites that can help increase your site's traffic.

Key Phrase- A phrase that is typed into a search engine to provide a reader with content related to the subject they are researching.

Keyword- A word typed into a search engine to provide a reader with content related to the subject they are researching.

Keyword Density- The percentage of keywords and phrases found within a web page versus the web page's full word count.

Link Building- Using SEM to link web pages to one another to promote website traffic and generate a profit.

Long Tail Phrase- A keyword phrase that consists of 3 or more words.

Meta Descriptions- The description provided on a search engine's results page to entice readers to click on content. Used for SERP's ranking.

Meta Tags- Keywords and phrases for SEO-- assist with generating website traffic and SERP's ranking.

Naked Links- Links that are not anchored within text. Appear in content as a URL.

Outbound Links- Links leading away from a website to a third party site.

Pay Per Clicks- Receiving payment for the number of times a link is clicked on by a reader.

Pay Per Views (PPV's)- Receiving payment for the traffic a web page receives.

Posting for Links- The method of posting links to gain visibility and website traffic.

Primary Keyword (Phrase)- The main keyword or phrase used for SEO

Reciprocal Links- Links used to connect two web pages to one another.

Search Engine Friendly- Content that can easily be found on SERP's.

Search Engine Marketing (SEM)- A method used to increase website traffic and profit.

Search Engine Marketing Professional Optimization (SEMPO)- The combination of SEO and SEM. Used to promote traffic and profit potential of websites.

Search Engine optimization (SEO)- Optimizing content to increase website traffic and ranking on SERP's.

Search Engine Results Page (SERP's)- The pages that appear on a search engine when a topic is searched.

Secondary Keyword (Phrase)- The second main keyword or phrase that is used with SEO.

User Generated Sites- Websites created by individuals that can lack credibility.

Word Count- The full number of words that are used to compose a piece of content.

About the Author

Alyssa Ast is an experienced freelance writer, who enjoys working from the comfort of her home office. Alyssa works for many online companies and private clients, fulfilling their needs for quality content. Using her SEO and SEM knowledge, Alyssa optimizes traffic to online content and websites-- increasing traffic and profit potential.

Although Alyssa thrives in the freelance writing world, her true passion is journalism. She is also an avid blogger. She currently blogs for the WM Network, as well as her personal blogs; Random Thoughts of a Tangled Mind and Writer's Block.

Alyssa is the co-founder of the WM Network, which she started with fellow freelance writer Angela Atkinson in May of 2009. This little side project has grown tremendously over the past year and continues to expand.

To give back to the writing community that has helped her so much, Alyssa offers her services as a freelance writer coach and SEO coach to aspiring writers.

Learn more about Alyssa Ast at www.alyssaast.com

www.ingramcontent.com/pod-product-compliance
Lightning Source LLC
Chambersburg PA
CBHW051217050326
40689CB00008B/1342